Disney's GOOFY VISITS PLUTO
A BOOK ABOUT THE PLANETS

Disney's GOOFY VISITS PLUTO
A BOOK ABOUT THE PLANETS

CATHY EAST DUBOWSKI

ILLUSTRATED BY
THE THOMPSON BROTHERS
AND
FRANC MATEU

Disney PRESS
NEW YORK

Copyright ©1995 by Disney Press.
All rights reserved.
No part of this book may be used or reproduced
in any manner whatsoever
without written permission from the publisher.
Printed in Singapore.
For information address Disney Press
114 Fifth Avenue, New York, New York 10011.

Consultant: Thomas W. Hocking
Education Coordinator
Morehead Planetarium
The University of North Carolina
Chapel Hill, NC

First Edition

1 3 5 7 9 10 8 6 4 2

Library of Congress Catalog Card Number: 94-71788
ISBN: 0-7868-3024-7

It was a beautiful, clear night in the neighborhood. "I can't wait to show Mickey my new telescope!" Goofy said. "It's a perfect night for looking at stars and planets!"

"Hi there, Goofy," said Mickey. "You're just in time."

"I am?" said Goofy.

"I'm on my way to visit Pluto," said Mickey.

"You *are*?"

"Wait here," said Mickey. "I'll go around back and get the car. Then we'll both go visit Pluto."

"We *will*?" said Goofy. "Gawrsh!"

Goofy was so excited. He and Mickey were going to visit Pluto! He sat down and rocked in the porch swing, staring dreamily into the sky. "I wonder which one of those cute little twinkly lights is the planet Pluto."

Suddenly Goofy felt very sleepy. "Boy, I sure hope Pluto isn't too far away," he said, yawning. "I can hardly keep my eyes open."

Beep beep!

Goofy rubbed his eyes. Then he shook his head and jumped to his feet. Mickey was in his car. But *was* it a car? "Gawrsh, Mickey," said Goofy, "your car looks just like a spaceship!"

"That's because it is a spaceship," said Mickey. "Hop in."

Goofy climbed in and buckled his seat belt.

"Welcome aboard!" crackled a voice from the control panel.

"Thanks!" said Goofy. "Hey, who said that?"

"That's PAL," said Mickey, "our computer copilot."

"Nice to meet you, PAL," greeted Goofy.

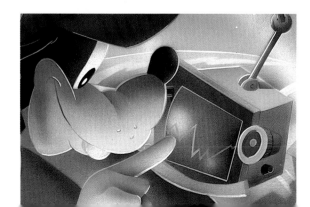

"Ready when you are, captain!" said Goofy.

Mickey grinned and started the countdown. "Ten . . . nine . . . eight . . . seven . . . six . . . five . . . four . . ."

The rockets roared to life. ". . . three . . . two . . . one . . .

"Blastoff!"

"Hey, Mickey," said Goofy. "How far away is Pluto, anyway?"

"Let's ask PAL!" said Mickey.

"Pluto is the planet farthest from the Sun, at a distance of nearly four billion miles."

"We better step on it then," said Goofy. "Mickey, can't you make this buggy go any faster?"

"Sure thing, Goofy!" said Mickey. He pressed a button on the control panel. "Hold on!"

Mickey and Goofy were pressed against their seats as the spaceship suddenly leaped forward and began hurtling through space.

"Whoaaaaaaaaa!" cried Mickey and Goofy. The ship was going faster and faster. Goofy closed his eyes and flung his arms around Mickey.

"I can't see where we're going!" yelled Mickey.

"Me, neither!" said Goofy.

Mickey and Goofy bounced up and down as the spaceship went cartwheeling out of control. "I feel lllike I'm in an outer space wwwashing mmmachine!" said Mickey.

"Mmme, tttoo!" said Goofy.

"Look!" said Goofy, pointing at a bright light just up ahead. "Somebody left a light on."

"Excuse me," said PAL. **"The spaceship is approaching the Sun. The Sun is a star. It is the star closest to Earth and therefore appears biggest and brightest. Its surface temperature is 10,000 degrees Fahrenheit. Translation: extremely hot! Recommend you step on brakes immediately!"**

Mickey slammed on the brakes. *Screech!* The spaceship slid to a stop.

"Now what?" Goofy asked PAL.

"TURN AROUND!"

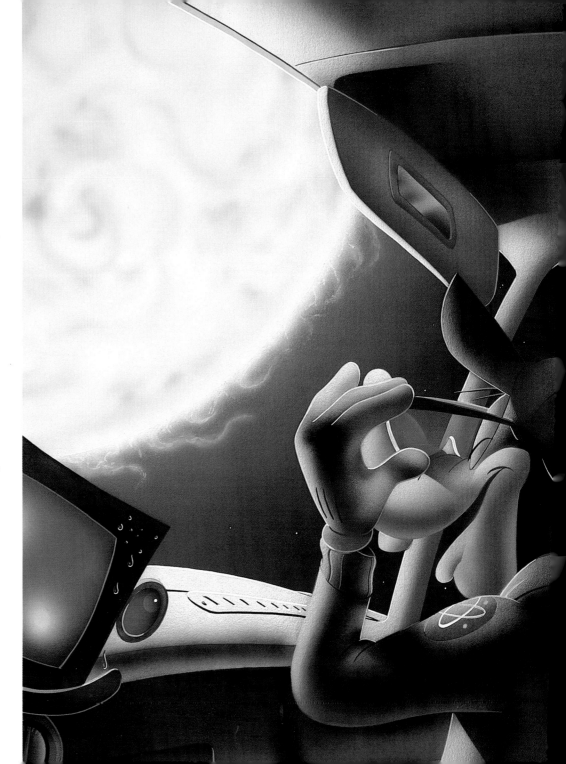

"Boy," said Mickey when the spaceship was turned around and out of danger. "That was a close call. Thank goodness for PAL."

"He sure is a pal!" said Goofy. "By the way, PAL," asked Goofy, "can you tell us how to get to Pluto?"

"There are nine planets in the solar system," said PAL. **"The first one we will come to is the planet Mercury."**

"I bet that's Mercury right there," said Mickey.

"Affirmative," said PAL. **"It is the planet closest to the Sun, and very hot. The temperature on the surface of Mercury can reach 800 degrees Fahrenheit."**

"Don't forget the suntan lotion!" said Goofy.

A little while later the rocket ship came upon a planet covered in clouds of gold and yellow.

"It's getting hot in here," said Goofy. "Can we turn on a fan?"

"Now approaching Venus," said PAL. **"It is the hottest planet in the solar system. The temperature on the surface is always around 900 degrees Fahrenheit."**

"How come it's the hottest planet if Mercury is closer to the Sun?" asked Mickey.

"The thick clouds on the planet's surface act like a lid to trap the Sun's heat," said PAL.

"The clouds are awfully pretty," said Mickey.

"The clouds are made up of poisonous gases," said PAL.

"Uh-oh!" said Goofy. "Better hold your nose!"

In the distance Goofy and Mickey saw a blue ball with brown patches and white swirls.

"I'd know that planet anywhere!" said Mickey. "It's Earth!"

"Hey, Mickey," said Goofy. "Do you think we can see our houses from up here?"

Mickey laughed. "I think we're too high up for that, Goofy."

"Your house is on the dark, nighttime side of Earth," explained PAL. **"The Earth spins like a top. On the side facing the Sun it is day; on the other it is night."**

"That means all our friends are sleeping," said Mickey.

"Ssshh," said Goofy. "We'd better not wake up Donald."

They stopped on the Moon so Mickey could take some snapshots for his photo album. Goofy went for a moon walk.

"Look at me!" said Goofy. "I'm bouncing like a ball."

"The Moon is much smaller than Earth," said PAL, **"so there is not as much gravity."**

"This would be a great place for a basketball game," said Goofy.

Back on course again, Goofy and Mickey came to a dry red planet covered with craters, mountains, and valleys.

"This is the planet Mars," said PAL, **"the fourth planet from the Sun."**

"I saw this planet in a movie," Goofy told Mickey. "It's where UFOs and little green men come from."

"Correction," said PAL. **"There are no little green men on Mars. It does have two tiny moons, however, named Phobos and Deimos."**

"Mickey has two tiny nephews," Goofy told PAL. "Their names are Morty and Ferdie."

Suddenly there was a big *klunk*, and the spaceship shook and rattled.

"PAL!" cried Mickey. "What's happening?"

"Do not be alarmed," said PAL. **"We are now passing through an asteroid belt."**

"An asteroid belt?" asked Goofy. "Gawrsh, I didn't know that asteroids even wore pants."

"They don't," PAL explained. **"Asteroids are chunks of rock and metal that orbit the Sun between Mars and Jupiter."**

"It's like a game of bumper cars," said Mickey as he steered through the asteroids.

Bonk!

"Only louder!" said Goofy.

Goofy and Mickey breathed a sigh of relief when the spaceship finally passed through the asteroid belt. Now they were in deep space. It was beginning to get very dark.

"We are now approaching what are known as the outer planets," said PAL.

"Golly!" said Mickey as the next planet came into view. "It's huge!"

"That is Jupiter," said PAL. **"It is the fifth planet from the Sun and the largest in the solar system. It is so big that all the other planets could fit inside it."**

"Look," said Goofy, "it's got a belly button!"

"That is what is called the Great Red Spot," said PAL. **"Scientists believe it may actually be a giant storm."**

As Mickey steered the spaceship past Jupiter, another beautiful planet came into view. It was almost as big as Jupiter and was encircled by seven brightly colored rings.

"We are now arriving at the planet Saturn," said PAL. **"Its rings are made of millions of pieces of ice and rock and dust. Saturn is the second biggest planet, and yet because it is mostly gas it is very light. If you could put it in a bathtub, it would float like a beach ball."**

"Who has a bathtub *that* big?" asked Mickey.

Mickey checked the speedometer on the control panel. "Gee, Goofy, we're already more than one and a half billion miles from home." Then he asked PAL, "What planet is that up ahead?"

"It's Uranus," said PAL, **"and like Saturn, it is a ringed planet."**

"It must be sleepy," said Goofy, trying not to yawn. "It looks as if it's lying down."

"Scientists think that something may have crashed into the planet and knocked it on its side," said PAL.

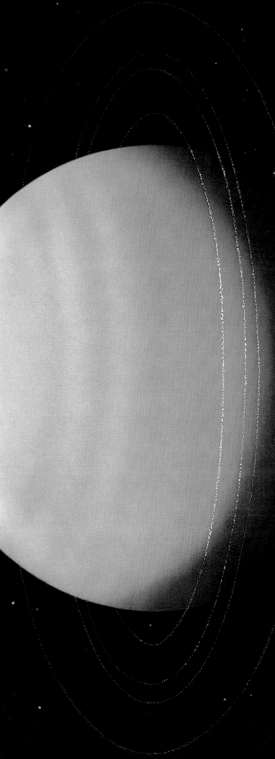

Goofy shivered as the spaceship traveled farther and farther into deep space. They were billions of miles from home now, and the blue planet that appeared before them looked cold and very lonely.

"The planet Neptune," announced PAL. **"Storms on the surface blow up to 700 miles an hour, making it the windiest planet in the solar system."**

"I wish I'd remembered to bring my kite," said Goofy.

"Neptune is the eighth planet from the Sun," said PAL. **"It is so far away from Earth that it can be seen only with a telescope."**

"Then it's a good thing I brought my new telescope along!" said Goofy.

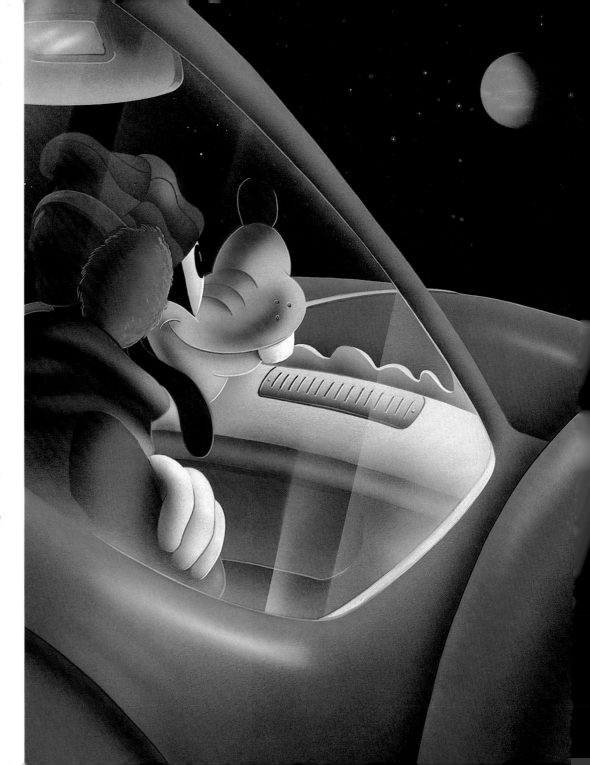

"The spaceship is now approaching the ninth planet in the solar system," said PAL.

"I know!" shouted Goofy. "Pluto!"

"Affirmative," said PAL.

"Gee," said Mickey. "Pluto sure is tiny."

"Pluto is the smallest planet in the solar system," said PAL. ***"It would take five Plutos to equal one Earth. In fact, scientists guess that Pluto may once have been a moon of Neptune."***

Goofy decided to take a space walk and make a video of their visit to Pluto.

"Don't forget your mittens and scarf," Mickey told him. "It's cold out there."

"That is correct," said PAL. **"The temperature on the surface of Pluto is 360 degrees below zero."**

"Hey, Mickey!" called Goofy as he floated through space. "This is more like a space *swim* than a space walk!"

Goofy wanted a closer look at Pluto. He jabbed at a button on his space suit. "I bet this will get me closer."

"Warning!" urged PAL. **"The jet pack has been activated."**

"Oops," said Goofy as the jet pack flared and he went spinning and tumbling through space. Goofy was racing straight for the surface of Pluto!

"Helllp!" he yelled.

"Goofy!" shouted Mickey. His voice seemed very far away. "Goooooofy!"

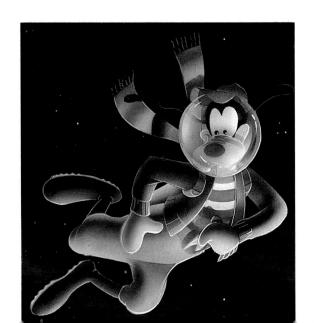

"Goofy!" Mickey shouted again. He honked his horn.
Beep beep!
Goofy's eyes blinked open. Mickey was sitting in his car . . . his very *ordinary* Earth car.
"Gawrsh," Goofy said, looking around. "What's going on? I thought we were visiting Pluto."
"We *are* visiting Pluto," said Mickey. "He's at the animal hospital with a sore paw. Come on, let's go."

"Pluto is fine," the veterinarian told Mickey. "He's ready to go home."

On the drive back, Goofy told Mickey all about his trip to Pluto.

"You must have been dreaming," said Mickey. "But we can still use my car to visit outer space. I'll show you."

Mickey drove all the way to the top of Lookout Hill. "Welcome to outer space," he said. Then he helped Goofy set up the telescope.

"Gawrsh!" said Goofy when he looked into the telescope. "You're right, Mickey! This *is* just like being in outer space. I can even see my footprints on the Moon!"

Mickey grinned and shook his head. "You must still be dreaming, Goofy," he said.

"Honest," said Goofy. "Look for yourself!"

PLUTO THE PLANET

- Pluto is a cold and distant planet.

- Pluto is 1,430 miles in diameter, less than one-fifth the width of Earth.

- Pluto is the smallest planet in the solar system.

- Pluto can be found 3.7 billion miles from the Sun.

- Pluto orbits the Sun once every 248 years.

- Pluto spins in an orbit that's *elliptical*—which means "oval shaped."

PLUTO THE DOG

- Pluto is a warm and friendly dog.

- Pluto is fourteen inches wide, twenty-six inches high, and weighs about ninety pounds.

- Pluto is one of the biggest dogs in the neighborhood.

- Pluto can usually be found snoozing in his doghouse.

- Pluto goes for a walk around the block with Mickey at least once a day.

- Pluto spins around chasing his tail—which means he gets very dizzy.

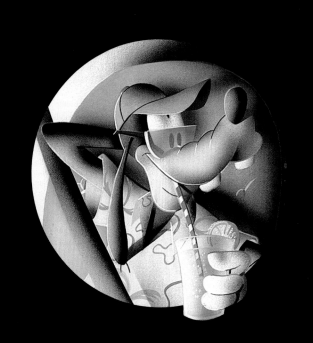